FARM ANIMALS

Featuring animals from

Gooseberry Bridge
FARM

Ways to use this book

Younger Readers

Read the book together and let your reader hear the clues and see the close-up picture as they guess the animal.

Older Readers

Read each clue aloud without showing the close-up picture. Let your reader try to guess the animal as you read the clues. Show the close-up picture if they need an extra clue.

Make It a Game!

Read the clues aloud and try to guess the animal in as few clues as possible. Come up with your own clues for each animal after reading the extra facts.

LEARNINGSPARKEDUCATIONALPUBLISHING

I eat **SEEDS** and **INSECTS** and sometimes **MICE** and **LIZARDS**.

I lay **EGGS**.

I can't fly **FAR**, but I can **FLY** into a **TREE** or over a **FENCE**.

There are over **25 BILLION** of me in the **WORLD**.

WHAT AM I?

I am a
CHICKEN.

CHICKEN FACTS

Chickens are a popular farm animal. In fact, there are over 25 billion chickens around the world. This means there are more chickens than any other type of bird in the world. It also makes them the most popular domesticated, or tamed, bird in the world.

There are hundreds of different breeds of chickens. Chickens come in all colors, patterns, and sizes. Some chickens are raised for meat. Other chickens are raised for eggs. Female chickens, called hens, lay eggs. Male chickens are called roosters.

Chickens are omnivores. This means they eat both plants and animals. Chickens usually eat grain provided by the farmer, but they will also eat insects, lizards, and even small mice.

Chickens aren't great at flying, but they can fly short distances. They can fly high enough to get over a fence or into a tree.

I have poor **EYESIGHT**.

I am one of the **SMARTEST** animals.

I use my snout to root for **FOOD**.

I am found all over the **WORLD**.

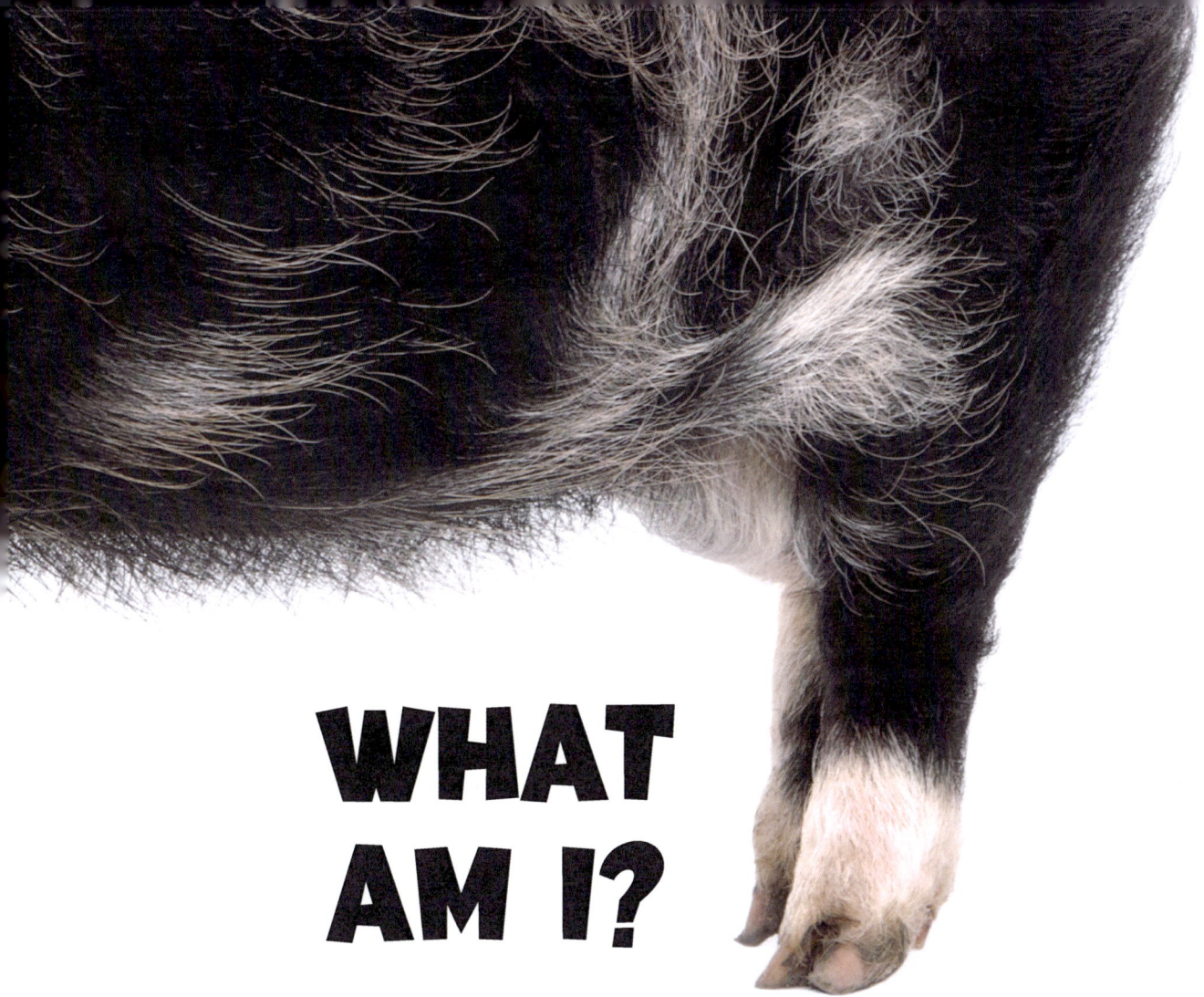

WHAT AM I?

I am a **PIG**.

PIG FACTS

Pigs are raised in many parts of the world. Pigs are usually raised for meat, but parts of the pig can be used for other things like paintbrushes, crayons, medicine, and more!

Pigs have poor eyesight but an excellent sense of smell. The pig uses its nose, called a snout, to search for food. Pigs search by rooting, or digging, with their snout. Pigs are omnivores, so they eat both plants and animals. Wild pigs eat leaves, roots, and even rodents or small reptiles. Pigs on farms eat a balanced diet provided by a farmer. This diet usually consists of corn and other grains.

Pigs have a reputation for being dirty animals. Pigs are actually very clean and only roll in mud to keep cool in hot weather. Pigs are also known for being smart animals. They are one of the smartest domesticated animals.

I am raised for **MEAT** and **MILK**.

I live in a **HERD**.

I **REGURGITATE** my food and chew it as **CUD**.

I am a **RUMINANT**.

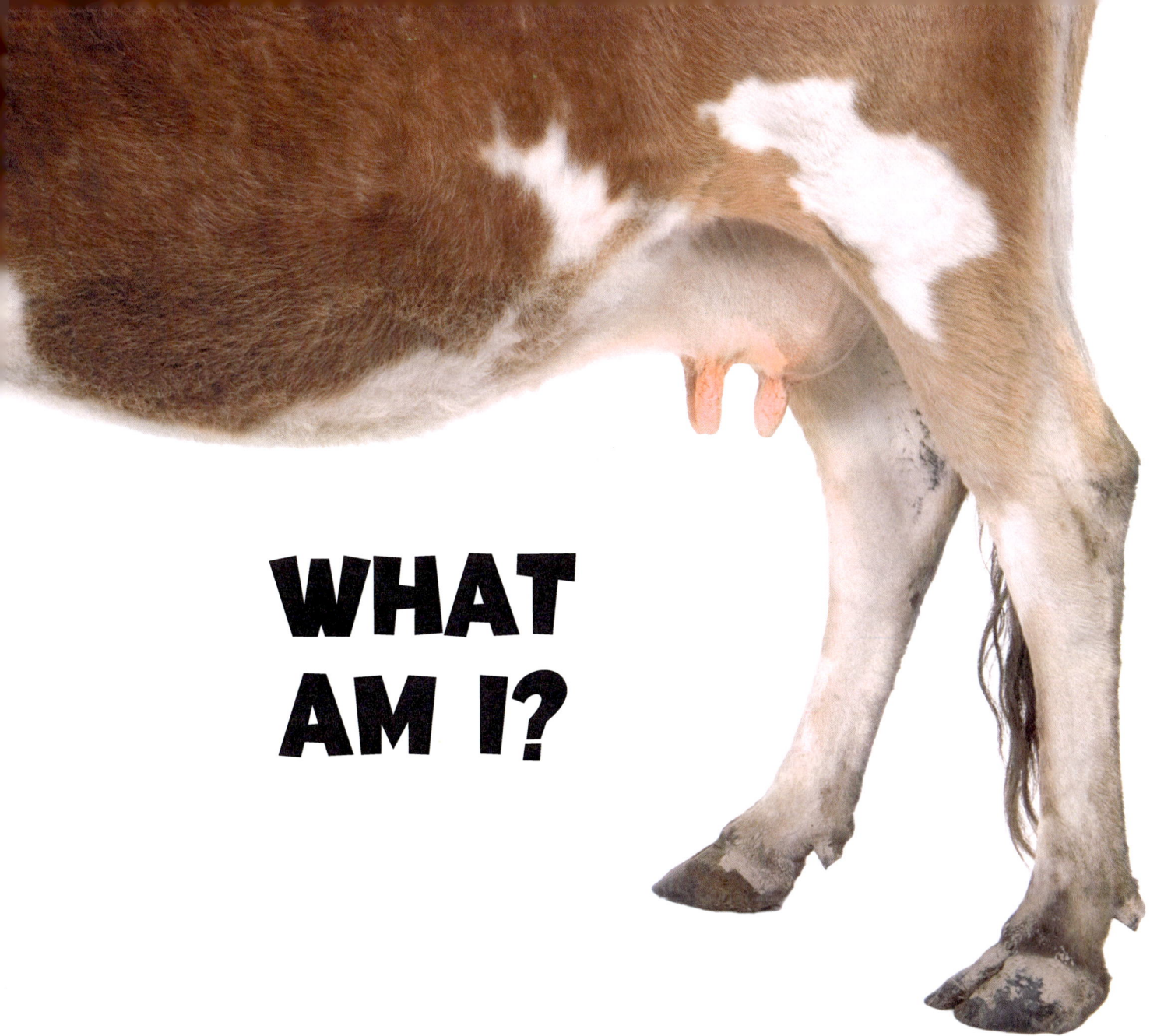

WHAT AM I?

I am a **COW**.

COW FACTS

The word "cow" is often used to refer to all cattle, but a cow is actually female and a bull is male. Cattle are raised all over the world. Some cattle are raised for meat, some for milk, and some for both meat and milk. Other products that come from cattle include leather, glue, buttons, makeup, medicine, and much more!

Cattle are ruminants. This means they have a single stomach with four compartments. The largest compartment in their stomach is called the rumen. Cattle regurgitate, or spit up, contents from the rumen to chew again. This regurgitated product is called cud. Cattle are herbivores, meaning they only eat plants. Some plants are difficult to digest. Chewing cud helps cattle digest all kinds of vegetation.

Cattle like to be around other cattle. They are social animals and live together in herds.

I am raised for **MEAT** and **WOOL**.

I am the only source of **LANOLIN**.

A **SHEPHERD** takes care of me.

I live in a group called a **FLOCK**.

WHAT AM I?

I am a **SHEEP**.

SHEEP FACTS

Sheep are raised all over the world. They are typically gentle animals and have little ability to defend against predators. Sheep rely on other sheep to group together in a flock for defense. Sheep also rely on humans to take care of them. People who take care of sheep are often called shepherds. Some flocks of sheep are very large. A shepherd often uses a herding dog to help protect and care for the sheep.

Sheep are raised for both wool and meat. Wool is the hair that covers a sheep's body. Sheep are usually trimmed, or shorn, once a year for their wool. Wool is naturally flame resistant and is used to make fabric and clothing. Wool also contains a type of grease called lanolin. Lanolin is used in makeup and lotions.

Sheep are herbivores and ruminants. They eat grass and other plants. They graze in areas that other farm animals can't.

My **TEETH** never stop **GROWING**.

Sometimes, I eat my own **POOP**.

I am raised for **MEAT** and **FUR**.

I have **LONG** ears and **STRONG** back legs.

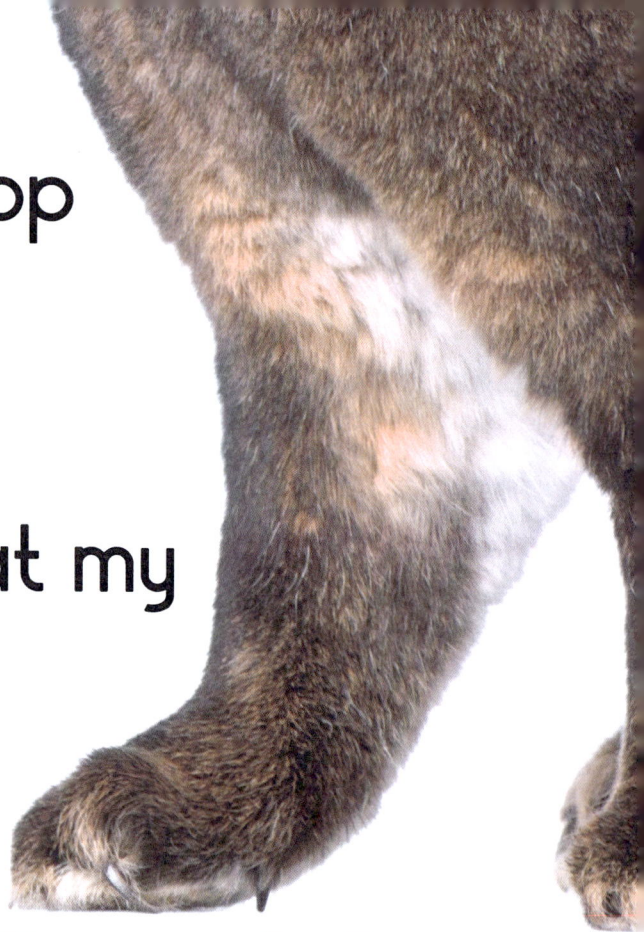

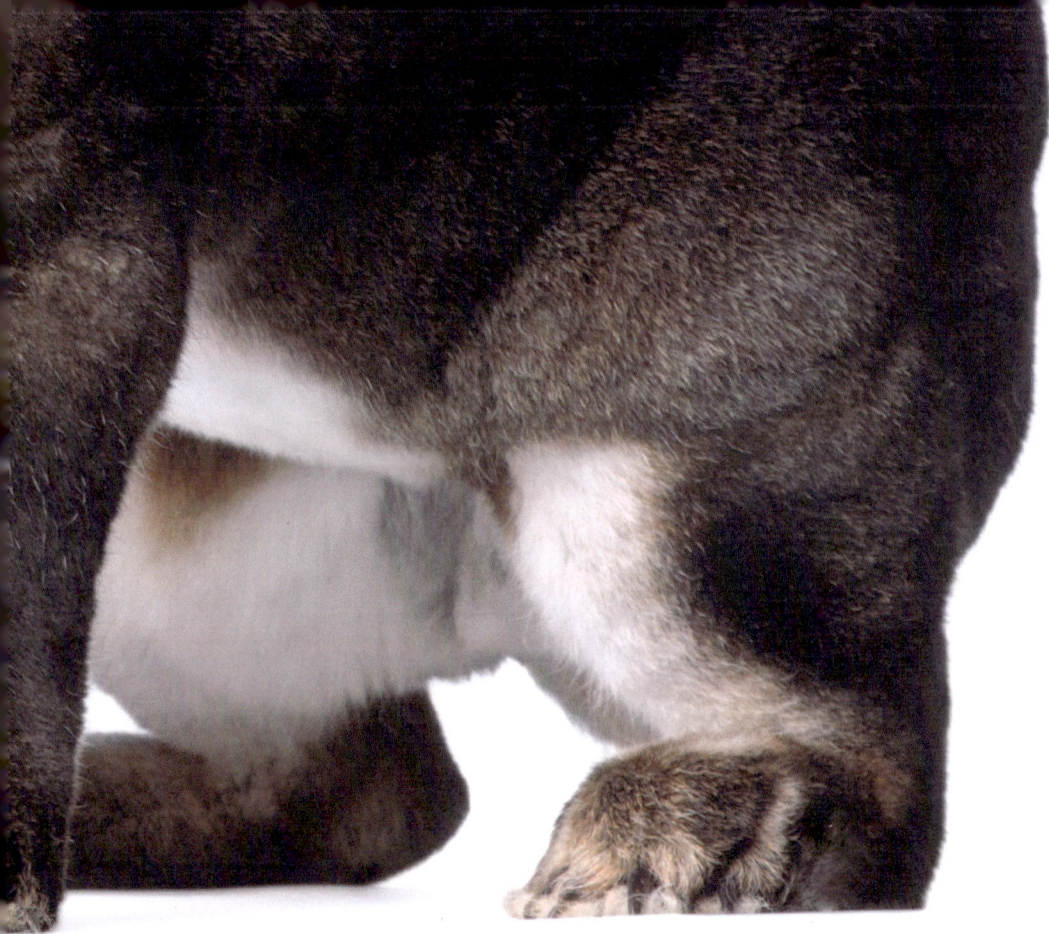

WHAT
AM I?

I am a **RABBIT**.

RABBIT FACTS

Rabbits are perfect for small farms. They take up little space and they are quiet. They are also typically gentle and easy to handle. Rabbits are raised for meat or fiber. Fiber is hair or fur.

Rabbits are herbivores. They eat small amounts of plants throughout the day. A rabbit's teeth are always growing, so it eats often to keep its teeth worn down. Rabbits also eat some of their own poop. This is called coprophagy. Rabbits produce two kinds of poop. Some is hard and some is softer. The softer poop is called cecotrope. Rabbits only eat the cecotropes which allows them to get more nutrients from the plants they eat.

Rabbits are known for their long ears and strong back legs. Their big ears help them hear well and stay cool in hot temperatures. They use their strong hind legs for hopping.

I am **EXCELLENT** at escaping enclosures.

I am raised for **MEAT**, **MILK**, and **FIBER**.

I am a **RUMINANT**.

I am one of the first **ANIMALS** to be **DOMESTICATED** by humans.

WHAT AM I?

I am a
GOAT.

GOAT FACTS

Goats are raised all over the world. They were one of the first animals to be domesticated by humans. They come in all sizes and colors. Some have horns and some are polled, meaning they don't have horns. Goats are raised for meat, milk, and fiber. More people around the world eat goat meat and drink goat milk than from any other farm animal.

Goats are ruminants. They have a stomach like a cow and chew cud too. They are herbivores and considered browsers. This means they eat different types of leaves, shrubs, and other things grazing animals like cattle won't eat. Goats like to eat plants from the top down. They don't like to eat close to the ground.

Goats are curious and agile animals. They like to climb on things and explore. They might jump a fence or figure out how to unlatch a gate. This makes them excellent at escaping enclosures.

I live in a **HIVE**.

I can fly up to
15 MILES per hour.

I am the only **INSECT** that
produces **FOOD** for humans.

Beating my wings makes
a **BUZZING SOUND**.

WHAT AM I?

I am a **HONEY BEE.**

HONEY BEE FACTS

Humans have cared for honey bees for thousands of years. Beekeeping, or apiculture, allows honey to be harvested without harming the bees or the hive. Some honey bees live in the wild, but most are domesticated. Most honey bees live in hives that are cared for by humans. Honey bees are the only insects that produce food for humans.

Honey bees live in a hive. There are three types of bees in a hive. The queen runs the whole hive and lays eggs. Workers are all female and have many different jobs. Some workers gather food from flowers, while others stay in the hive to protect it and care for the eggs and baby bees. Drones are all male and help the queen produce eggs.

Honey bees visit 50 to 100 flowers in one trip from the hive. They fly up to 15 miles per hour. They beat their wings more than 200 times per second. This rapid wing movement makes the buzzing sound associated with honey bees.

Clues by YOU!

It's your turn to create clues. Look at the pictures and read the facts for the following farm animals. Can you come up with four clues for each animal based on their appearance, behavior, or some other fact?

Tell your clues to a friend or family member and see if they can guess the animal without looking at the picture.

I am an ALPACA.

Alpacas are native to the Andes in South America. Alpacas were domesticated a long time ago, and none are found in the wild.

Alpacas are raised mostly for wool but also for meat in some parts of the world. Alpacas live in a herd and like to be with other alpacas. They are the smallest members of the camel family.

Alpacas are herbivores and ruminants. They can survive on less food than other animals of a similar size.

I am a COTURNIX QUAIL.

The Coturnix quail is a game bird, but it is also raised in a farm setting. It is raised for eggs and meat and grows quickly. Due to its small size, it is perfect for small farms or even back yards.

Quail make their nests on the ground. They eat seeds and grains, but they also eat insects and worms. Coturnix quail are quiet and calm.

I am a DOMESTIC TURKEY.

Domestic turkeys are found throughout the world. They are native to North America and were domesticated in Mexico hundreds of years ago. Turkeys are raised for meat.

Domestic turkeys share the same name as wild turkeys, but they are very different. Domestic turkeys are usually white, while wild turkeys are dark brown. Wild turkeys are also much smarter than domestic turkeys.

With special thanks to Gooseberry Bridge Farm
for their photogenic animals.

Gooseberry Bridge
FARM

To learn more about Gooseberry Bridge Farm
and see more cute animals, check them out on
Instagram or Youtube.
@gooseberrybridgefarm

Or visit gooseberrybridgefarm.com

Text copyright © 2024 Kizzi Roberts

Photographs © Kizzi Roberts; Staci Hill; Jeremy Hill; lifeonwhite/elements.envato.com;
fotorince/elements.envato.com; cynoclub/elements.envato.com.

Published in May 2024 by Learning Spark Educational Publishing in Rogersville, Missouri.
Learning Spark Educational Publishing is an imprint of Elemental Ink Publishing LLC.

Library of Congress Control Number: 2024910263

Hardcover: 979-8-88884-032-0; Paperback: 979-8-88884-031-3

Edited by Carrie Rodell. Book design and layout by Kizzi Roberts.

www.LearningSpark.com

www.ingramcontent.com/pod-product-compliance
Lightning Source LLC
Chambersburg PA
CBRC090247230326
41458CB00108B/6515